George William Gray

The chlorides of paranitroorthosulphobenzoic acid

George William Gray

The chlorides of paranitroorthosulphobenzoic acid

ISBN/EAN: 9783337305987

Printed in Europe, USA, Canada, Australia, Japan

Cover: Foto ©berggeist007 / pixelio.de

More available books at www.hansebooks.com

Dissertation

Submitted to the Board of University Studies of the Johns Hopkins University for the Degree of Doctor of Philosophy.

By

G. Wm. Gray.

1895

Contents

Acknowledgement		5
I. Introduction		6
II. Preparation of Material		9
III. Action of Phosphorus Pentachloride upon the acid Potassium salt of paranitro orthosulphobenzoic acid		12
Analyses of the Chlorides		14
IV. Action of Water on the Chlorides		20
V. Action of Ammonia on the Chlorides		21
VI. Isomeric salts		26
	a. The Silver Salts	26
	b. The Potassium Salts	29
	c. The Barium Salts	31
	d. The Calcium Salts	34
	e. The Magnesium Salts	36
	f. The Zinc Salts	39
	g. The Ethereal Salts	41
VII. Paranitro cyanbenzenesulphonic acid		44

VIII	Paranitrocyanbenzene sulphuric chloride	117
IX	Paranitro cyanbenzene sulphonamide	48
X	Paranitro cyanbenzene sulphon amide	50
XI	Action of Aniline on the Chlorides of Paranitro orthosulph benzoic acid	51
	a. On the unsymmetrical chloride in the Presence of water	51
	b. On the dry unsymmetrical chloride	52
	c. On the Unsym Chloride in Chloroform Solution	52
	d On the Symmetrical chloride	54
XII	Infusible Amide	55
XIII	Fusible Amide	57
XIIII	The Anul	55
XV	Aniline Salt of Paranitro. the sulphanilidobenzoic acid	60
XVI	Action of Paraluidin on the chlorides	61
	a. Paranitro orthosulphobenzo chloride (Inua)	61
	b. Paranitro orthosulphbenzo Chloi. (Inua)	63
XVII	Conclusion	64
XVIII	Biographical	68

Acknowledgment

This work was suggested by Professor Remsen and was carried out under his direct supervision. To him I desire to express my sincere thanks for the instruction received in the laboratory as well as in the lecture room. I also wish to express my obligation to Professors A. N. Morse and W. B. Clark of the Geological Dept.

1 Introduction.

In the course of an investigation on the action of aniline upon the chloride of orthosulphobenzoic acid Remsen and Coates[1] in 1891 obtained two isomeric amlides with quite different properties, and an aml. This suggested that the chloride was not an individual. In the following year Remsen and Kohler[2] obtained fairly conclusive evidence of the existence of two chlorides, one of which they obtained as a pure crystalline product, the other as an oil. The crystallized chloride on treatment with aniline

[1] Amer. Chem. J. / ... p 311
[2] " " " " 330

yielded only two products the fusible anilide and the acid, the oil yielded besides these two products an infusible anilide. Remsen and Saunders[1] the next year, obtained both chlorides in crystalline form.

In consequence of this work it was deemed advisable to study the action of phosphorus pentachloride upon paranitroothosulphobenzoic acid and see if it would yield isomeric chlorides. The results which I have obtained and which are given in detail in the following pages show that two chlorides are obtained, though Kastle[2] describes only one as the product of such action.

In this article I shall speak of

1 J. a. Chem. Jour. V. 17 P. 347
2 Amer. Jour. V.

the two chlorides as the symmetrical and the unsymmetrical. The evidence which leads to this view is obtained from the action of ammonia upon the two chlorides. When the unsymmetrical chloride is treated with ammonia, it is converted into the ammonium salt of para-nitrocyanbenzenesulphonic acid

$$C_6H_3\begin{cases}CN\\SO_2ONH_4\\NO_2\end{cases}$$

and this compound can be converted into a chloride by treatment with phosphorus pentachloride

$$C_6H_3\begin{cases}CN\\SO_2Cl\\NO_2\end{cases};$$

on the other hand the symmetrical chloride gives with

... a ... ... salt
of paranitrobenzacsulphimide.
This ammonium salt

$$C_6H_3\left\{\begin{matrix}CO\\SO_2\end{matrix}\right\rangle N.NH_4$$
$$NO_2$$

in treatment with acids gives
paranitrobenzoic sulphimide, which
crystallizes in fine needles and
melts at 209°

II Preparation of Material

100 grams of paranitrotoluene were
mixed with 400 grams of fuming sulphuric acid and heated
in a tall flask at a tem-
perature of 100°. by being immersed in boiling water. The mixture
was kept at this temperature un-

til a portion of it, on being poured into water, gives a nearly clear solution. This heating usually requires from 6 to 8 hours. The mixture is then allowed to cool and poured into cold water. After standing for several hours, it is filtered and then neutralized with calcium carbonate. The calcium salt is filtered from the calcium sulphate and converted into the potassium salt by treatment with potassium carbonate. The potassium salt is filtered from the calcium carbonate and evaporated to crystallization. 100 grams of the potassium salt of paramidothene sulphonic acid, together with 30

grams of potassium hydrate are dissolved in 500 cc of water; the latter flask containing this solution is immersed in water and heated to a temperature of 105°. When the solution has acquired this temperature, 220 grams of potassium permanganate are added, and the mixture heated until the color of the permanganate has entirely disappeared. The amber colored liquid is then filtered from the manganese hydrate, and the dark brown residue thoroughly washed with hot water. The filtrate and washings are then neutralized with hydrochloric acid, and evaporated until the neutral potassium salt begins

to crystallize. & when a small quantity of the solution is cooled the whole solution is then allowed to cool and strong hydrochloric acid added in excess, when the acid potassium salt of paranitro orthosulphobenzoic acid separates out in beautiful fine white needles.

III Action of Phosphorus Pentachloride upon the Acid Potassium Salt of Paranitroorthosulphobenzoic Acid

When one molecule of the anhydrous acid potassium salt of paranitroorthosulphobenzoic acid is intimately mixed with two molecules of phosphorus pentachloride

no action or only very slight action takes place in the cold. 40 grams of the acid potassium salt are mixed with 60 grams of phosphorus pentachloride and heated in a distilling bulb at a temperature of 150°. The distilling bulb is immersed in a sulphuric acid bath and kept at a temperature of 150° for four or five hours. Reaction takes place with the formation of two chlorides:

$$C_6H_3\begin{cases}COOH\\SO_2OK\\NO_2\end{cases} + 2PCl_5 = C_6H_3\begin{cases}CCl_2\\SO_2O\\NO_2\end{cases} + 2POCl_3 + HCl + KCl$$

$$C_6H_3\begin{cases}COOH\\SO_2OK\\NO_2\end{cases} + 2PCl_5 = C_6H_3\begin{cases}COCl\\SO_2Cl\\NO_2\end{cases} + 2POCl_3 + HCl + KCl$$

The yield of the first is from 80-90% and of the second from 10-20%. The chlorides must not be heated

much above 150° as decomposition begins at 170°. The yield of the second chloride is increased to about 30% by heating in an open dish upon the water-bath. The oily mass is poured into water and washed free from phosphorus oxychloride and potassium chloride. The chlorides are then dissolved in chloroform, the chloroform solution dried by means of calcium chloride; after filtering off the calcium chloride, the solution is allowed to evaporate slowly in a bell-jar free from moist air. From this solution a considerable quantity of the solid chloride (symmetrical) is obtained in crystals which look as if they consisted of monoclinic(?)

pina cords and basal frame. These crystals are filtered off and washed with ice-cold chloroform. After no more crystals can be obtained by this method, more chloroform is added and ~~the~~ solution transferred to an Erlenmeyer flask. This flask is immersed in a freezing-mixture of salt and ice and dry air drawn over the solution for several days. From this solution more crystals of the symmetrical chloride are obtained. These are filtered off and washed very quickly with cold chloroform. The chloroform solution was evaporated until all the chloroform had been driven off, and the oil was then dissolved in a large volume of petroleum ether

(boiling point 55°–90°) to the solution cools, it becomes cloudy and then begins to clear up. When this happens pour the solution into another beaker and allow the solution to continue to cool. From this solution the unsymmetrical chloride then crystallizes almost pure.

This (unsymmetrical) chloride is slightly soluble in petroleum ether 1 pt in 200, and crystallizes from this solvent in fine needles or long flat plates about 3 to 4 mm broad and 2 to 3 cm long. They are extremely soluble in ether and chloroform, but crystals can not be obtained from these solvents. It is decomposed on heating above 160°

and melts at 56-57 (uncor.) This chloride is probably identical with that obtained by Kastle[1].

The symmetrical chloride can be obtained pure by recrystallizing the first crystals obtained above. It can also be obtained by dissolving the two chlorides in chloroform and shaking this solution with a little dilute ammonia. The ammonia will act immediately upon the unsymmetrical chloride, and only slightly upon the symmetrical. The chloroform solution is washed free from ammonia, and dried by means of calcium chloride and allowed to evaporate. From the chloroform solution beautiful crystals are obtained which ap-

pear to consist of a combination of monoclinic prisma coids and basal planes. It is also soluble in ether but less so than the iso meric chloride. It melts at 94-95° (Anschütz) It can be heated in a sealed tube with phosphorus oxychloride to a temperature of 130° without undergoing any decomposition.

Both chlorides when in crystalline form, can be left exposed to moist air without undergoing very much decomposition. When the chlorides are first formed, before crystals are obtained, decomposition goes on more readily judging from the amount of hydrochloric acid given off.

Analyses of the Chlorides

Portions of both chlorides were heated in a sealed tube with fuming nitric acid. The results obtained were as follows:

0.1733 gr Unsym. chloride gave 0.1763 g. AgCl.
0.1933 " " " " 0.1545 BaSO₄
0.1927 " " " " 0.1930 AgCl
0.1727 " " " " 0.1548 BaSO₄

Calculated for $C_6H_3\{^{CCl_2}_{^{SO_2}_{NO_2}}}$		Found	
		I	II
Cl	24.97	25.01	25.02
S	11.25	11.25	11.06

The symmetrical chloride on analysis gave
0.1901 gr sym. chloride gave 0.1935 AgCl
0.1850 " " " 0.1875 AgCl

Calculated for $C_6H_3\{^{COCl}_{^{SO_2Cl}_{NO_2}}}$		Found	
		I	II
Cl	24.97	25.15	25.06

IV. Action of Water on the Chlorides

Cold water acts very slowly upon the two chlorides; even after several months the decomposition is incomplete. In hot water they are more readily decomposed, but on boiling the decomposition is complete in a few minutes. The decomposition is about as rapid in one case as in another. With dilute caustic potash they are readily decomposed and after cooling an excess of hydrochloric acid was added when the characteristic acid potassium salt of paranitroorthosulphobenzoic acid crystallized out in fine needles thus showing that both yield the same decomposition product.

Action of Ammonia upon the Chloride

Kastle¹ mentions a product of the action of ammonia upon the chloride which he obtained but gives no description of it. It was therefore thought best to repeat this work.

5 grams of the symmetrical chloride were dissolved in ether and poured into a beaker containing 150 cc of dilute ammonia. The solution was allowed to stand for several days until the ether evaporated. From this solution the ammonium salt of paradichlor urea sulphinic was obtained. The yield is theoretical. The crystals are rectangular plates or glistening flakes

... ... it
boiling point of sulphur was
It is also soluble in alcohol. On
adding hydrochloric acid to this
salt paramid. benzoic sulphamide
is obtained. This sulphamide crys
tallizes in plates or fine needles
and melts at 209° (Ulmer).

$C_6H_3 \begin{cases} CO\ Cl \\ SO_2\ Cl \\ NO_2 \end{cases} + 4\ NH_4OH = C_6H_3 \begin{cases} CO \\ SO_2 \end{cases} N\ NH_4 + 2\ NH_4Cl + 4H_2O$

$C_6H_3 \begin{cases} CO \\ SO_2 \\ NO_2 \end{cases} N\ NH_4 + HCl \quad C_6H_3 \begin{cases} CO \\ SO_2 \\ NO_2 \end{cases} N\cdot H\ ,\ +\ NH_4Cl.$

When the perfectly dry chloride
is dissolved in anhydrous ether
and dry ammonia gas passed
into the solution the same pro
uct is obtained.
0.3066 grams of this salt gave 0.5 ...
0.3000   "   "   "   0.5178 " "

Calculated for		Found	
$C_6H_3\{{}^{CO}_{SO_2}\}NNH_4$			
	NO_2		
N	17.18	17.16	17.26

5 grams of the unsymmetrical chloride were dissolved in ether and poured into a beaker containing 100 cc of dilute ammonia. After the ether had evaporated nothing crystallized from the solution until it had evaporated to about 50 cc when a large mass of needles were obtained. These needles grouped themselves in radiating masses, or tufts. This product proved to be the ammonium salt of para nitrocyanbenzene sulphonic acid. This salt is also obtained when the mixed chlorides are dis-

solved in chloroform and shaken with dilute ammonia. From the ammoniacal solution the ammonium salt is obtained on evaporation. From the chloroform solution the symmetrical chloride is obtained. When ammonia is poured upon the unsymmetrical chloride solution takes place immediately. In the case of the symmetrical chloride solution is not complete even after 24 hours.

$$C_6H_3\begin{cases}CCl_2\\SO_2\\NO_2\end{cases} + 4NH_4OH = C_6H_3\begin{cases}CN\\SO_2NH_4\\NO_2\end{cases} + 2NH_4Cl + 4H_2O$$

It is probable that an unstable diamid

$$C_6H_3\begin{cases}C(NH_2)_2\\SO_2\\NO_2\end{cases}$$

is fact, ... that the ... by a ... molecular change passes over into the ammonium salt of paramidobenzenesulphonic acid.

$$C_6H_3 \begin{cases} NH_2 \\ SO_2 \cdot ONH_4 \\ NO_2 \end{cases}$$

The analysis of this salt gave the following results:

On standing over sulphuric acid 0.9965 grams lost .0687 grams H_2O.

Calculated for
$$C_6H_3 \begin{cases} NH_2 \\ SO_2 \cdot ONH_4 + H_2O \\ NO_2 \end{cases} \qquad Found$$

H_2O 6.85   6.90

0.3000 grams of the anhydrous salt gave 0.05189 N
0.3... " " " " 0.05196 "

Calculated for
$$C_6H_3 \begin{cases} NH_2 \\ SO_2 \cdot ONH_4 \\ NO_2 \end{cases} \qquad Found$$

N   17.15   17.16   17.32

All [...] of [analyses?] which
are given in grams were made
by the Kjeldahl method, all others
were by the absolute method.
On adding acid to this salt
nothing separates out even on
boiling. As further study was de-
sirable in order to place the formation of this salt beyond doubt,
a number of salts derived from
these two ammonium salts were
made.

VI. <u>Isomeric Salts</u>.
a. <u>The Silver salts</u>.

Silver [P]aranitrocyanbenzenesulphonate
$C_6H_3 \begin{cases} CN \\ SO_2OAg + H_2O \\ NO_2 \end{cases}$ — This salt was
prepared by treating ammonium

paramite oxantingenas with the nick silver nitrate. To the h.t lution of the ammonium salt silver nitrate is added, and the solution allowed to cool. On cooling the silver salt crystallizes in fine long rectangular needles 2 cm long. This salt contains one mol ecule of water of crystallization 0.1744 gram of this salt lost .0334 gram of H_2O at 100°.

Calculated for
$C_6H_3\{^{CN}_{SO_2\cdot OAg}_{NO_2}+H_2O$ Found

H_2O 5.23   5.10

The anhydrous salt on analysis gave the following results.

0.1779 grams gave .0710 gram $AgCl$
0.1556 "   "   .0667 "   "
0.2010 "   "   15.1 cc N

Calculated for
$C_6H_3\begin{Bmatrix}Cl\\SO_2\cdot Ag\\NO_2\end{Bmatrix}$ Found

		I	II
Cl_1	32.21	32.11	32.26
A	8.35	8.13	

Silver salt of the Sulphinide $C_6H_3\begin{Bmatrix}CO\\SO_2\end{Bmatrix}N\cdot Ag$
NO_2

This salt was prepared by Noyes'
by adding silver nitrate to the ammonium salt of the sulphinide.
It crystallizes in short fine needles,
difficultly soluble in boiling water,
insoluble in cold water. It contains
no water of crystallization. It
turns brown on exposure to the
light.

0.1508 gram gave .0645 gram AgCl
0.1995   "   "   .0856   "   "

Calculated /
$C_6H_3 \begin{cases} CN \\ SO_2 \text{\textgreater} N.Ag \\ NO_2 \end{cases}$ Found

                                I   II

Ag   32.21   32.20   32.30

/ The Potassium Salt

Potassium Paranitrocyanbenzenesulphonate
$C_6H_3 \begin{cases} CN \\ SO_2OK + 1½ H_2O. \\ NO_2 \end{cases}$ — This salt is prepared by treating the silver salt with sufficient potassium chloride to precipitate all the silver. The solution is filtered from the silver chloride and evaporated to crystallization. The potassium salt crystallizes in beautiful colorless needles 2 to 3 cm in length and contains 1½ molecules of water of crystalli-

gation.

0.3177 gram of this salt lost .0287 gram H_2O at 110°

$$C_6H_3\begin{cases}Cl\\SO_2K\cdot 1\tfrac{1}{2}H_2O\\NO_2\end{cases}$$

Calculated for   Found

H_2O 9.21   9.03

0.1868 gram of the anhydrous salt gave .0605 gram K_2SO_4

0.2495 gram of the anhydrous salt gave .0802 gr K_2SO_4

calculated for

$$C_6H_3\begin{cases}Cl\\SO_2K\\NO_2\end{cases}$$

         Found

        I   II

K   14.61   14.54   14.43

<u>Potassium salt of the Sulthinate¹</u>

$$C_6H_3\begin{cases}CO\\SO_2\end{cases}\!\!A\cdot K$$
$$NO_2$$

This salt is formed

1. Amer chem Jour   "

ly heat, paranitrobenzenesulphonate with potassium carbonate until neutral reaction. It crystallizes in fine leaflets and contains no water of crystallization. It is readily soluble in hot water but difficultly soluble in cold.

0.2352 grm of this salt gave 0.757 gram K_2SO_4

	Calculated for $C_6H_3\begin{cases}SO_2\mathord{\cdot} K\\NO_2\end{cases}$	Found
K	14.66	14.45

c. *The Barium Salts.*

Barium Paranitrocyanbenzenesulphonate

$\left(C_6H_3\begin{cases}CN\\SO_2O\\NO_2\end{cases}\right)_2 Ba + 2\tfrac{1}{2} H_2O$ — This salt is made by treating the silver salt with ba-

... chloride. It could not be obtained in well formed crystals, but formed warty like masses. It is readily soluble in cold water and contains 2½ molecules of water of crystallization.

0.3747 gram of this salt lost 0.0263 gram H₂O
0.3698 " " " "   gave 0.1325 gram BaSO₄

Calculated for
$$\left(C_6H_3\left\{\begin{matrix}CH_3\\SO_3\cdot\\NO_2\end{matrix}\right\}_2\right)Ba + 2\tfrac{1}{2}H_2O$$
                                Found

H_2O 7.07   7.07
Ba 21.05   21.23

It loses 1½ molecules of water of crystallization over sulphuric acid.
0.2073 gram anhydrous salt gave 0.0822 gram BaSO₄.

Calculated for
$$\left(C_6H_3\left\{\begin{matrix}CH_3\\SO_3\cdot\\NO_2\end{matrix}\right\}_2\right)Ba$$
                                Found
Ba 23.17   23.09

Barium Salt of the Sulphinide $\left(C_6H_3\left\{\begin{matrix}CO\\SO_2\end{matrix}\right\}N\right)_2Ba + 3H_2O$

This salt is obtained by heating the sulphinide with barium carbonate. The solution is filtered from the excess of barium carbonate and allowed to crystallize. The salt forms well developed prisms which contain 3 molecules of water of crystallization.

0.3881 gram of this salt lost 0.0329 gram H_2O on…

Calculated for
$$\left(C_6H_3\left\{\begin{matrix}CO\\SO_2\\NO_2\end{matrix}\right\}N\right)_2Ba + 3H_2O$$

                                    Found

H_2O 8.37   8.414

0.3552 gram of the anhydrous salt gave 0.1396 gram $BaSO_4$.

Calculated for
$$\left(C_6H_3\left\{\begin{matrix}CO\\SO_2\\NO_2\end{matrix}\right\}N\right)_2Ba$$

                                    Found

Ba   23.17   23.10

d. **The Calcium Salts.**

Calcium Paranitrocyanbenzenesulphonate
$(C_6H_3\{^{CN}_{SO_2O}\}_{NO_2})_2 Ca + 7H_2O$ — This salt is formed by heating the silver salt with calcium chloride. It, like all the salts of the cyanbenzenesulphonic acid, is extremely soluble in water. It is in fact soluble in less than its own weight of water. It forms flat plates which are elongated. It loses 3 molecules of water of crystalization over sulphuric acid and four more at a temperature of 135°.

0.6459 gram of this salt lost 0.0541 gram H_2O over H_2SO_4 sulphuric acid

0.6459 gram of this salt lost 0.1294 gram H_2O at 135°

Calculated for:
$(C_6H_3\{^{CN}_{SO_2O}\}_{NO_2})_2 Ca + 7H_2O$ Found

$7H_2O$	20.31	20.04

0.2129 gram of the anhydrous salt gave
0.0661 gram $CaSO_4$.

Calculated for

$$\left(C_6H_3\left\{\begin{matrix}CI\\NO_2\end{matrix}\right\}\right)_2 Ca$$

Ca	8.07	8.10

Calcium Salt of the tryptham $i.e.$ $C_6H_3\left\{\begin{matrix}Cl\\SO_3\\NO_2\end{matrix}\right\}_2 Ca + 6H_2O$

This salt is formed by heating paranitro-benzenesulphonide with calcium carbonate. It forms long transparent, rect-angular columns which soon lose their transparency and become opaque, and lose part of their water of crystallization.

0.7069 gram of this salt lost 1272 gram water at 100°.

Calculated

$$\left(C_6H_3\left\{{CO \atop SO_2}{>A \atop NO_2}\right\}\right)_2 Ca + 6H_2O \qquad \text{Found}$$

$6H_2O \qquad 17.43 \qquad\qquad 17.49$

0.1913 gram anhydrous salt gave 0.0526 gram $CaSO_4$

$$\text{Calculated for} \qquad\qquad \text{Found}$$
$$\left(C_6H_3\left\{{CO \atop SO_2}{>A \atop NO_2}\right\}\right)_2 Ca$$

$Ca \qquad 8.09 \qquad\qquad 8.09$

e. <u>The Magnesium Salts.</u>

Magnesium Paranitrocyanbenzenesulphonate.
$C_6H_3\left\{{CN \atop SO_2O \atop NO_2}\right\}_2 Mg + 8H_2O$ — This salt is formed by treating the silver salt with Magnesium Chloride. It crystallizes in prisms which are somewhat elongated. It is extremely soluble in

water ... ... 8 ... ...
water of crystallization, of which it
loses 3½ over sulphuric acid
·3500 gram(?) this salt lost ·...74 gram H_2O ... H_2SO_4
 "   "   0.0842   "   "   at 2 ...

Calculated for
$(C_6H_3\{{}^{CH_3}_{SO_2C}_{NO_2}\})_2 Mg \cdot 18H_2O$ Found

3½ H_2O —   13.01   13.00
S   "   23.12   23.21

0.1693 gram of the anhydrous salt gave ·411 gram $Mg_2P_2O_7$
0.7116 "   "   "   "   "   0.0493 "   "

Calculated for
$C_6H_3\{{}^{CH_3}_{SO_2C}_{NO_2}\}_2 Mg$ Found
                                               /   II

Mg   5.09   5.31   5.09

Magnesium salt of the sulphon. $(C_6H_3\{{}^{CH_3}_{SO_2C}_{NO_2}\})_2 Mg + 6H_2O$

This salt is ??? by heating a ??? ??? sulphamide with magnesium carbonate. The salt forms beautiful transparent crystals which seem to be a combination of orthorhombic domes and prisms ???. These crystals soon lose their lustre and become opaque. The salt contains 6½ molecules of water of crystallization.

0.2672 gram of this salt lost 0.0527 gram H_2O
1.2934 "   "   "   "   " 0.2570 "   "

Calculated for
$$(C_6H_3\{{}^{CO}_{SO_2}{}^{>N}_{}\}_2 Mg + 6\tfrac{1}{2} H_2O)$$

Found
                           I   II
6½ H_2O 19.64   19.60   19.87

0.2165 gram of the anhydrous salt gave 0.0523 g MgO

Calculated for
$$(C_6H_3\{{}^{CO}_{SO_2}{}^{>N}_{}\}_2 Mg)$$

Found
Mg   5.119   5.??

The Zinc Salt

Zinc Paranitrocyanbenzenesulphonate
$(C_6H_3\{{}^{CN}_{SO_2O}_{NO_2}\})_2 Zn + 7H_2O$ — This salt is formed when the Zinc salt is heated with zinc chloride. It forms long prismatic columns, is extremely soluble in water and contains 7 molecules of water of crystallization. 0.7453 gram of this salt lost 0.1477 gram H_2O at 170°.

Calculated for
$(C_6H_3\{{}^{CN}_{SO_2O}_{NO_2}\})_2 Zn + 7H_2O$ Found

$7H_2O$ 19.52   19.52

0.2064 gram of the anhydrous salt gave 0.0318 gram ZnO
0.1710   "   "   "   "   "   "   0.0264 "

Calculated for   Found
$(C_6H_3\{{}^{CN}_{SO_2O}_{NO_2}\})_2 Zn$
Zn   12.60   12.38   12.48

Zinc salt of the Sulphimide, $C_6H_3\{{SO_2 \atop NO_2}\}N\}_2 Zn + 4\frac{1}{2}H_2O$

This salt is formed by treating paranitrobenzoesulphimide with zinc carbonate. It forms short thick prisms combined with domes, and resembles very much the corresponding magnesium salt. It contains $4\frac{1}{2}$ molecules of water of crystallization.

0.1861 gram of this salt lost 0.0254 gram H_2O at 125°
0.4327 " " " 0.0593 " " "

Calculated for
$(C_6H_3\{{SO_2 \atop NO_2}\}N)_2 Zn + 4\frac{1}{2}H_2O$ Found

$4\frac{1}{2} H_2O$ 13.49   13.65   13.71

0.2102 gram of the anhydrous salt gave 0.0328 gram Zn.

Calculated for
$(C_6H_3\{{CO \atop SO_2}\}N)_2 Zn$ Found
^{NO_2}

Zn 12.60   12.54

4. Ethereal Salts

The Methyl Salt of the Sulphimide $C_6H_3\{{}^{CO}_{SO_2}\}N.CH_3$ —
This salt is formed by heating the silver salt of paramidobenzoic sulphimide in a sealed tube, at 100° with methyl iodide.

$$C_6H_3\{{}^{CO}_{SO_2}\}^{NO_2}N.Ag + CH_3I = C_6H_3\{{}^{CO}_{SO_2}\}^{NO_2}N.CH_3 + AgI.$$

The methyl salt is extracted with boiling alcohol. On cooling the salt crystallizes in leaf like crystals. It is readily soluble in hot alcohol but difficultly soluble in cold alcohol; it is also soluble in water. It melts at 179° (uncorr.)

0.1575 gram of the salt gave 14.73 cc N.
0.1561 "   "   "   "   " 14.25 " "

Calculated for
$C_6H_3\{{}^{CO}_{SO_2}\}^{NO_2}N.CH_3$ Found

N   11.57   11.77   11.47

From the product of the action of methyl iodide upon the silver salt of nitrometacyanbenzene sulphuric acid nothing definite has been obtained.

The Ethyl Salt of the Sulphimide

$$C_6H_3 \left\{ \begin{array}{l} SO_2 \\ NO_2 \end{array} \right\} N.C_2H_5 \; —$$

This salt is formed when the silver salt of the sulphimide is heated in a sealed tube at 100° with ethyl iodide.

$$C_6H_3 \left\{ \begin{array}{l} SO_2 \\ NO_2 \end{array} \right\} N.Ag + C_2H_5 I = C_6H_3 \left\{ \begin{array}{l} SO_2 \\ NO_2 \end{array} \right\} N.C_2H_5 + AgI.$$

The mass is extracted with boiling alcohol, filtered and allowed to cool, when a mass of needles resembling dinitrobenzene is obtained. It is readily soluble in hot alcohol but difficultly soluble in cold, it crystallises

in 1 liter. It melts at 172° Uncor

0.1309 gram of this salt gave 13.53 cc N.
0.1645   "   "   "   "   "   14.68   "

Calculated for
$C_6H_3\{\begin{matrix}CN\\SO_2\end{matrix}>N.C_2H_5$ Found
                                            I   II

N.   10.97   11.26   11.21

From the silver salt of isocyanbenzene sulphonic acid I have been unable to obtain an ethereal salt by the action of ethyl iodide, but have obtained a substance which is strongly acid, the nature of which has not yet been determined, but is probably analogous to the orthobenzaminesulphonic acid obtained by Jeswein by the action of alcohol on orthocyanbenzene sulphonic acid

VII. Paramidoyanbenzenesulphuric Acid
$C_6H_3\{{}^{CN}_{SO_2OH}{}_{NO_2}, H_2O$ — This acid is formed when its silver salt is treated with hydrochloric acid enough to precipitate all the silver. It is also formed when the barium salt is precipitated with sulphuric acid. The acid crystallizes in long prisms which contain one molecule of water of crystallization. If heated very rapidly, it melts between 145° and 150° but solidifies again immediately. This is due to the ~~

fact that it acts yet in solution in its water of crystallization and when this is driven off it solidifies. It is extremely soluble in water, being soluble in less than its own weight. On being neutralized with potassium carbonate the potassium salt of paranitroxy-anbenzenesulphonic is obtained. On boiling the acid with hydrochloric acid, the acid ammonium salt of paranitrouthosulphobenzoic acid is probably obtained for when an excess of caustic potash is added ammonia is liberated, and then on adding an excess of hydrochloric acid a mass of fine needles is obtained resembling the acid potassium salt of paraamidothosulphobenzoic acid.

When the acid is heated at 130°
it loses one molecule of water and
if the temperature is then raised to
150° it begins to increase in weight
This increase is probably due to the
oxidation of the cyanogen group.
This product is still under investigation.

0.3603 gram of the acid lost .0234 gram H_2O .1130
0.2303 "   "   "   "   gave .02161 grams A.
0.3033 "   "   "   "   .03403 "   "

$$\text{Calculated for}$$
$$CH_3 \begin{cases} C \cdot N \\ SO_2OH \\ NO_2 \end{cases} + H_2O \qquad \text{Found}$$

H_2O 7.31   7.06
A 11.42   11.43   11.23

VIII. Paramidocyanosulphobenzenchloride $C_6H_3\begin{cases} CN \\ SO_2Cl \\ NO_2 \end{cases}$ —

This chloride is formed when the potassium salt of the paramido-cyanbenzenesulphonic acid is heated with phosphorus pentachloride. One molecule of the potassium salt is mixed with one molecule of phosphorus pentachloride and the mixture heated to a temperature of 140° in a sulphuric-acid bath for several hours. The mass is then treated with water to decompose the phosphorus oxychloride and to dissolve the potassium chloride. The chloride is then dissolved in chloroform and dried by means of calcium chloride. The chloroform solution is allowed to evaporate at the

chloride crystallizes in long wet angular prisms which melt at 107°–108° Uncor. They are soluble in Chloroform, ether and benzene.

0.2640 gram of this Chloride gave 0.2338 gram BaSO₄
0.1640   "   "   "   0.1372 "   "
0.2025   "   "   "   18.75 cc N.
0.1640   "   "   "   0.0984 grams AgCl
0.2640   "   "   "   0.1544 "   "
0.1003   "   "   "   0.0592 "   "

Calculated for
$C_6H_3 \begin{cases} Cl \\ SO_2Cl \\ NO_2 \end{cases}$ Found

Cl   14.35   14.55   14.46   14.59
N   11.40   11.63
S   13.00   13.23   13.19

II. Paranitrocyanbenzene sulphonamide
$C_6H_3 \begin{cases} Cl \\ SO_2NH_2 \\ NO_2 \end{cases}$ — This amide is formed

When pure benzene sulfamide(?) which is treated with ammonia 2 grains of the chloride were dissolved in chloroform and an aqueous solution of ammonia added. The mixture is shaken in a separating funnel occasionally. The amide separates as an amorphous mass, which is filtered off and recrystallized from alcohol. It crystallized in small rectangular blocks resembling cane sugar. It does not melt below 270° yield excellent. 0.1920 grm of the salt gave 28.02 c.c.
0.1535 " " " " 22.79 "

    Calculated for
        $C_6H_7\begin{cases}CN\\SO_2NH_2\\N_2\end{cases}$ Found
  N.   18.33   18.54   18.65

X. Para aminoxybenzsulfonanilide $C_6H_5\{{SO_2NHC_6H_5 \atop NH_2}\}$

This anilide is formed in the same way as the anilide discussed from by the chloride in chloroform and add aniline, then cover the mass with water and heat the mixture to drive off the chloroform. As the chloroform is evaporated the anilide separates as a powder. This powder is recrystallized from alcohol. The yield in this case as in that of the amid is theoretical. The anilide crystallizes in fine needles and melts at 207°-208° (Uncor).

0.1513 grams of this anilide gave 16.97 cc N.

Calculated for
$C_6H_5\{{SO_2NHC_6H_5 \atop NH_2}\}$

N. 13.94   Found
                              14.09

XI. Action of Aniline on the Chlorides of Paraorthosulphotoxic Acid

" Action of Aniline on the Unsymmetrical Chloride in presence of Water. The unsymmetrical chloride was added to an excess of aniline in and emulsion in water 4 gram of chloride in 30 cc of water. The mixture was shaken until no more aniline was taken up. The excess of aniline was then distilled off in a current of steam. The liquid was removed by filtration and from the filtrate aniline hydrochloride and the aniline salt of paraorthosulph-p-amidobenzoic acid were obtained. The residue insoluble in water was dried and extracted with chloroform. The chloroform on evaporation gave

nearly pure fourth anilide of paranichodithosulphetongeor acid. The residue from the chloroform ~~extract~~ was dissolved in ~~chloroform~~ alcohol and was the infusible anilide. The yield of the infusible was only about 30%, the fusible anilide about 20% and the aniline salt about 30%.

b.— Action of aniline on the dry Unsymmetrical Chloride — ½ gram of the unsymmetrical chloride was gradually mixed with an excess of aniline. The mixture after being thoroughly ground, was heated on the water bath for 3 hours and was treated as above. There were obtained .4 gram of the fusible anilide and less than .1 gram of the infusible.

c. Action of Aniline upon the Unsym-

unsymmetrical Chloride   Chloroform Solution

.5 gram of the unsymmetrical chloride was dissolved in chloroform and an excess of aniline added to the solution. The solution was heated to drive off the chloroform and then heated at 100° for an hour. The mass was then treated as a.p.51. From the water solution .47 gm. aniline salt of paranitroorthosulphanilidobenzoic acid, and .23 gram of the fusible anilide. In another experiment in which I used the last of the chloride, the chloride was dissolved in chloroform and aniline added in excess. The mixture was shaken for a short time. The precipitate which formed was filtered and washed with chloroform; the precipitate was found insoluble anilide. The filtrate was

heated and from it consisted of the insoluble anilide and aniline hydrochloride. The chloroform solution was then heated and from it the fusible anilide and a small quantity of the infusible anilide were obtained.

d. *Action of Aniline on the Symmetrical Chloride* —, When the symmetrical chloride is heated with an excess of aniline in water, chloroform or ether the only products obtained are the fusible anilide and aniline hydrochloride. If the chloride is mixed with aniline and heated in the proportion of one molecule of chloride to one of aniline, the anil is formed; if in the proportion of one molecule of the chloride to two molecules of aniline the anil is formed together with a

a very small amount of the fusible anilide, when the proportion is one molecule of the chloride to four molecules of aniline only the fusible anilide is obtained

__Infusible Anilide__ — This is formed best by treating the unsymmetrical chloride dissolved in chloroform with an excess of aniline. The solution must be kept cold as the yield of this anilide is diminished by warming the solution. The infusible anilide is precipitated together with aniline hydrochloride in the chloroform solution. This precipitate is dried, and washed with water to dissolve the aniline hydrochloride, and the residue is found to be pure infusible anilide. It is insoluble in water. Soluble in alcohol. (Some recent

seems to me that this anilide undergoes a change by boiling it in an alcoholic solution. This is still under investigation. The product analysed was the above precipitate insoluble in water. It does not melt on heating to 250°. It is soluble in caustic potash but differs from the infusible anilide of orthosulphobenzoic acid in that it is decomposed by it. This substance on analysis gave the following results.

0.1735 gram infusible anilide gave 0.01826 grams
0.1818 " " " " 0.019221 " "
0.1925 " " " " 0.1156 " BaSO₄
0.1561 " " " " 0.0907 " "

$$C_6H_4 \begin{cases} C(NHC_6H_5)_2 \\ SO_2 \end{cases} > 0$$

	calculated for	found	
N	10.58	10.53	10.58
S	8.06	8.25	7.95

It was impossible to get it to crystallize except as wart like masses

<u>Fusible Anilide</u> This is formed in all cases (in part) in which and excess of aniline is used. It is the only product formed when an excess of aniline acts on the symmetrical chloride. In one experiment in which some infusible anilide was boiled in alcohol it was found to have been changed into the fusible anilide. This is still under investigation. The anilide crystallizes from chloroform in a silky mass. It is soluble in alcohol, insoluble in water. It is dissolved by alkalies and reprecipitated by acids; by this means it can be obtained free from the anil and the infusible anilide. It melts at $222°$ (uncor).

This salt on analysis gave the following results

0.2435 gram fused anilite ... 0.5158 grams CO_2
0.2453 "   0.0424 " H_2O
0.1502 "   0.0199 " $BaSO_4$
0.2016 "   17.4 cc   N
0.2035   17.50 "   N

$C_6H_5 \begin{cases} CO\, NH\, C_6H_5 \\ SO_2\, NH\, C_6H_5 \\ \end{cases}$ Calculated for

	I	II	III	IIII
C	57.40	57.34		
H	3.78	3.45		
S	8.06	8.21		
N	10.61	—	10.54	10.80

XVIII **Di Anil** — This is formed when aniline and the symmetrical chloride are heated together in the proportion of one molecule to of the Chloride to one molecule of aniline. Add the chloride gradually to the aniline and heat the mixture some time in a

sulphuric acid bath. The mass is then extracted with boiling alcohol. From the alcoholic solution the amil is obtained. It crystallizes in fine needles: it is insoluble in alkalies and by this means it can be separated from the amides. It melts at 183° (uncor). On analysis it gave the following results:

0.1562 gram of the amil gave 0.1199 gram $BaSO_4$
0.1735 "   "   "   0.1314 "   "
0.2105 "   "   "   15.82 cc N.
0.2124 "   "   "   15.67 "   "

Calculated for

$C_6H_3 \begin{cases} SO_2 \\ NO_2 \end{cases} N.\, C_6H_5$

		Found	
S	10.52	10.55	10.41
N	9.21	9.44	9.27

x

XI

Aniline salt of Paranitroorthosulphanilide-benzoic Acid — This is the fourth aniline derivative formed by the action of aniline upon the chloride of paranitroorthosulphobenzoic acid. Its formation is doubtless due to some secondary reactions. It is readily soluble in water, from which it crystallizes in fine flakes; on analysis it gave the following results

0.2664 gram gave 0.395 gram CO₂ & 0.1029 gram H₂O
0.2156 "   " 0.4411 "   " & 0.0856 "   "
0.1428 "   " 0.0787 "   BaSO₄
0.1434 "   " 0.0793 "   "
0.2000 "   16.36 cc N
0.2004 "   16.80 cc N

Calculated for

$$C_6H_3 \begin{cases} COOH_3 C_6H_3 \\ SO_2 NH C_6H_3 \\ NO_2 \end{cases}$$ Found

C	54.91	55.23	55.03	—	—
H	4.10	4.29	4.35		
N	10.13			10.40	10.55
S	7.71	—	—	7.60	7.58

XI. Action of Paratoluidin & the chlorides— The symmetrical and unsymmetrical chlorides were heated with in exactly the same way with paratoluidin. On using an excess of paratoluidin the same product was obtained from both chlorides

(a) Paranitroorthosulpho benzo toluide (trans) This is formed by treating either the symmetrical or the unsym-

methical chloride with para toluedin in chloroform solution or grinding them up together. It crystallizes from alcohol in fine slightly yellow needles which melt at 234° (Uncor). It is readily soluble in hot alcohol difficultly soluble in cold alcohol. It is dissolved by alkalies and reprecipitated by acids. It was found impossible to make any compound analogous to the infusible anilide. On analyses it gave the following results

0.2447 gram   gave 0.4433 gram CO₂ and 0.1019 gram H₂O
0.2576 "   "   0.4617 "   "   0.1077 " "
0.2022 "   " 16.46 cc N.
0.2078 "   " 16.97 " "
0.1611 "   "   0.0463 gr Ba SO₄
0.1405 "   "   0.1008 " "

Calculated for
$C_6H_3\begin{cases}CO\cdot NH\cdot C_6H_4\cdot CH_3\\SO_2\cdot NH\cdot C_6H_4\cdot CH_3\\ \end{cases}$ Found

	I	II	III	IIII
C	59.27	59.34	59.47	
H	4.48	4.54	4.47	
N	9.91			10.22 ; 10.26
S	7.53			7.37 7.67

Para toluidine symmetric toluol (para) –

b. Para Tluil. This is formed by heating the symmetrical Chloride with para toluidine in the proportion of one molecule of the Chloride to one molecule of the toluidine. The mixture is heated in a sulphuric acid bath for about an hour. The mass is then extracted with boiling alcohol. On cooling the tluil crystallizes from the alcohol in fine needles which melt at 164°.165° (Uncor). It is soluble in hot alcohol; insoluble in water and alkalies

On analyses it gave the following
results

0.1042 gram substance gave 0.0741 gram $BaSO_4$
0.1061   "   "   0.0756   "   "
0.1062   "   7.55 c.c.
0.1063   "   7.67   "   "

Calculated for
$C_6H_3 \begin{cases} CO \\ SO_2 \end{cases} N \cdot C_6H_4 \cdot CH_3$

                                    % Found
S   10.06   10.04   10.19
N   8.83   8.93   8.99

XIII Conclusion

Several conclusions seem to
be justified by the results of the
experiments described in this
paper
1. When phosphorus pentachloride

acts upon para nitroortho sulpho benzoic acid two isomeric chlorides are formed, one of which has the formula $C_6H_3 \begin{Bmatrix} C(Cl_2) \\ SO_2 \\ NO_2 \end{Bmatrix}$ and melts at 56–57° (Uncorr), the other has the formula $C_6H_3 \begin{Bmatrix} COCl \\ SO_2Cl \\ NO_2 \end{Bmatrix}$ and melts at 94–95° (Uncorr).

II. When the two chlorides are treated with ammonia, the first or unsymmetrical gives the ammonium salt of paranitro cyanbenzene sulphonic acid; the other gives the ammonium salt of paranitro benzoic sulphimide.

III. When the silver salts are treated with hydrochloric acid the silver salt of paranitro cyan benzene sulphonic acid gives the free ~~cyan~~ paranitro cyan benzene sulphonic acid, while

the silver salt of the sulphinide gives paraamidobenzene sulphinide

IIII When the silver salts are heated with methyl and ethyl iodide, the silver salts of the sulphinide yield ethereal salts while the silver salt of the paranitrocyanbenzene sulphonic acid yields a product which is strongly acid. The nature of this product has not yet been determined

V. When the two chlorides are heated with aniline the symmetrical chloride yields an amid and a fusible amlide while the unsymmetrical yields an infusible amlide which seems to be easily decomposed. It also yields the fusible amilide.

VI With paratoluidin both chlorides

yield the same toluide.

XLIII Biographical

George William Gray, the author of this dissertation, was born at Cumberland Md. on August 12, 1867. He was educated in the public schools of Maryland and Washington, D.C. He entered Johns Hopkins University in 1887 and graduated from there in 1890 having pursued the chemical-physical course. From 1890-1892 he was with the Baltimore Sugar Refinery and the Maryland Steel Company as chemist. He returned to Johns Hopkins University in the fall of 1892 and since that time has pursued advanced courses in chemistry.

www.ingramcontent.com/pod-product-compliance
Lightning Source LLC
Chambersburg PA
CBHW022134160426
43197CB00009B/1286